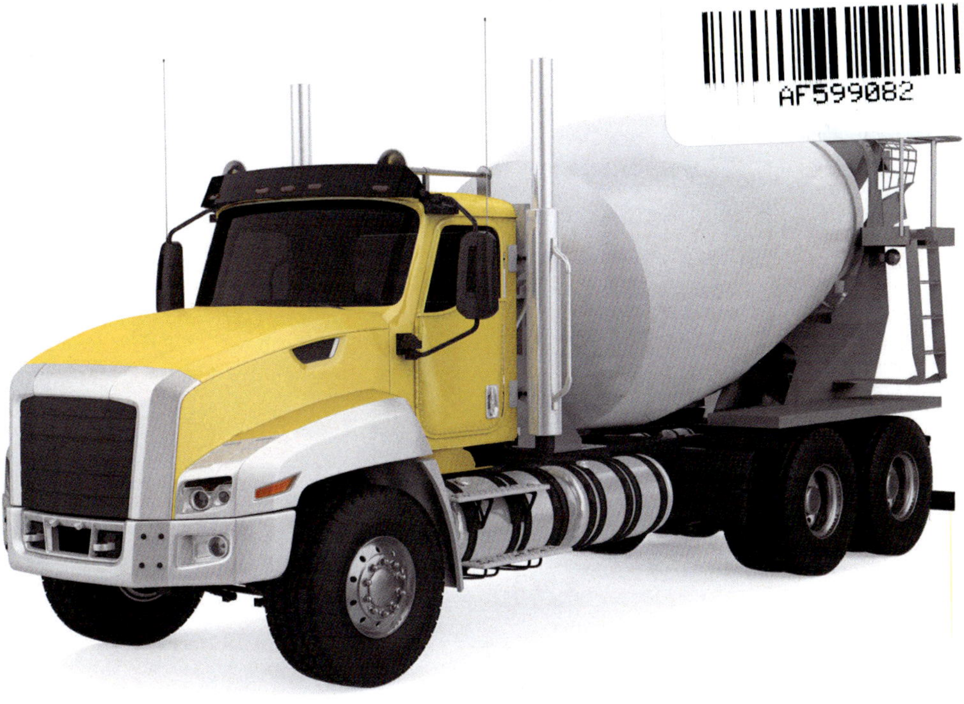

A Crabtree Roots Book

Crabtree Publishing
crabtreebooks.com

School-to-Home Support for Caregivers and Teachers

This book helps children grow by letting them practice reading. Here are a few guiding questions to help the reader with building his or her comprehension skills. Possible answers appear here in red.

Before Reading:

- What do I think this book is about?
    - *I think this book is about how concrete mixers work.*
    - *I think this book is about how concrete mixers are made.*
- What do I want to learn about this topic?
    - *I want to learn what concrete mixers do.*
    - *I want to learn the parts of a concrete mixer.*

During Reading:

- I wonder why...
    - *I wonder why some jobs need a concrete mixer.*
    - *I wonder what concrete is made of.*
- What have I learned so far?
    - *I have learned that many concrete mixers have six wheels.*
    - *I have learned that all concrete mixers make concrete.*

After Reading:

- What details did I learn about this topic?
    - *I have learned that some concrete mixers are orange.*
    - *I have learned that all concrete mixers have a drum.*
- Read the book again and look for the vocabulary words.
    - *I see the word **wheels** on page 6 and the word **drum** on page 10. The other vocabulary words are found on page 14.*

This is a **concrete mixer.**

Some concrete mixers are orange.

LIEBHERR

Many concrete mixers have six **wheels**.

All concrete mixers make **concrete**.

All concrete mixers have a **drum**.

How many concrete mixers do you see?

Word List

Sight Words

a	how	see
all	is	six
are	make	some
do	many	this
have	orange	you

Words to Know

concrete

concrete mixer

drum

wheels

34 Words

This is a **concrete mixer**.

Some concrete mixers are orange.

Many concrete mixers have six **wheels**.

All concrete mixers make **concrete**.

All concrete mixers have a **drum**.

How many concrete mixers do you see?

Written by: Ryan James
Designed by: Rhea Wallace
Series Development: James Earley
Proofreader: Melissa Boyce
Educational Consultant: Marie Lemke M.Ed.

Photographs:
Shutterstock: Wojciech Wrzesien: cover; Nerthuz: p. 1; Aaron M: p. 3; Victorio Shapiro: p. 5; Blanscape: p. 6-7; Sakoat contributor: p. 9; Another77; Margargee Films: p. 13

Crabtree Publishing

crabtreebooks.com 800-387-7650

Printed in Canada/012024/CP20231127

Published in Canada
Crabtree Publishing
616 Welland Ave.
St. Catharines, Ontario
L2M 5V6

Published in the United States
Crabtree Publishing
347 Fifth Ave
Suite 1402-145
New York, NY 10016

Library and Archives Canada Cataloguing in Publication
Available at Library and Archives Canada

Library of Congress Cataloging-in-Publication Data
Available at the Library of Congress

Hardcover: 978-1-0398-3838-3
Paperback: 978-1-0398-3923-6
Ebook (pdf): 978-1-0398-4006-5
Epub: 978-1-0398-4078-2